Ample Time

by Lorie Ann Hicks

IMOCO Publishing
Tulsa, OK
USA

Ample Time

by Lorie Hicks

Published by
IMOCO Publishing
P.O. Box 471721
Tulsa, OK 74147-1721

email: imoco@aol.com

ISBN 0-9661165-0-X

Library of Congress Catalog Number:
00-102132

"Most of the time management concepts I've studied over a 30-year career only work for a small percentage of people; I believe reading and implementing the Ample Time system will be effective for everyone, regardless of personalities, management styles, or habits."

— T.D. "Pete" Churchwell, President
Public Service Company of Oklahoma

"Ample Time speaks the truth...and helps in prioritizing what is important in life...on what you are trying to Be."

— Paula Marshall-Chapman, CEO/President,
The Bama Companies, Inc.

"Ample Time is simple, quick, and works for everyone. Lorie has skills most of us dearly wish we possessed. This book is her way of sharing and helping those of us challenged by hectic schedules."

— Karen Keith, News Anchor & Producer
KJRH, Oklahoma Living

"Reading this book was like having a conversation with a friend and coach. When you are ready to be productive and not just busy... implement Ample Time in your life. It is a simple, effective & quick way of making the most of life!"

— Joyce Powell, President
YGM Consulting Company

"The brevity of this book is a real plus and your insight refreshing. Most of us want a quick and effective way to manage our time and I commend you for finding 'ample time' to assist those of us in need."

— Linda Bradshaw, President
Tulsa World of Gymnastics

***Lorie Hicks** is the author of Ample Time. Lorie owns Corporate Challenges Consulting, A Marketing & Motivational Solutions Company, and teaches on the subject of time management both in workshops for her company and for Tulsa Community College. She has won numerous awards for her organizational skills & management, including an award by the Metropolitan Tulsa Chamber of Commerce as Tulsa's Young Entrepreneur of the Year.*

Visit www.corporatechallenges.com for more information.

This book is dedicated
to several very special people in my life.

To my mother who is the most beautiful person I know.
A sincere thank you for your unending belief in me and for being a great mentor of how to successfully live & appreciate every moment.

Thanks to my loving and supportive husband, Justin.
You are my anchor, my balance, my friend and a true partner.

To Lisa Holloway who first encouraged me to teach on time management and who continues to offer me friendship and support.

To my "Nan" for friendship in perfect form.

To Deanne & Cecy for being a great cheering section since our days at TWU.

To my friend, Lynn Martin.
Your miraculous experiences have reminded me that every moment counts!

To my brother, J.B., whom I love, adore and admire & his awesome wife, Leslie.
Thanks for being my sunshine and friends through the years.

To my publisher, Paul Brodsky, whom I've been lucky to know for many years.
You are an inspiration and a true friend. Thank you for your guidance and support.

To Vicki Young, who is first my friend and secondly, my assistant.
You are the kind of person everyone wants to have on their team.
Lucky, lucky me!

Many thanks to Steven Roop for cover design and Cynthia Reed for page design.

To all my family members, dear friends and clients
who have allowed me to experience life with them.
I cherish the memories and count my blessings
for the experiences and friendships
that have brought me to the incredible life
that I am so very privileged to live today.

Thank you, all.

Lorie

TABLE OF CONTENTS

INTRODUCTION

I don't really know more about time management than you do. I just seem to have a God-given gift for helping people "pull it together" and remember what they already know. When things get out of control, my customers, friends and family call me to gain insight on how to get re-focused, re-organized and re-energized. I'm going to spend these next few pages sharing this God-given gift of mine with you, simply reminding you of some principles of living a full life that you've forgotten. The principles of my Ample Time System are simple, yet powerful, and address absolutely EVERY possible "time challenger" you might be facing. When you apply the principles of time management that I share with you to your own lifestyle and needs, I guarantee you will experience days full of Ample Time for ALL of the things that are important to you. This is your opportunity to ditch the easy excuse of, "there just aren't enough hours in my day," and move forward to the most fulfilling and rewarding lifestyle you can imagine. I have some good news for you. There are enough hours in your day for everything you want to accomplish. The bad news is, even you are not so important that there is more in the world for you to complete in a day, a week, a month, a year or a lifetime than is physically possible. But before we can accomplish the most important goals in our lives, we must first discover our true priorities.

Do you believe that everything that has happened in your life, seemingly good and bad, prepares you for the very challenges and opportunities you are experiencing now? Thanks to my mother, I've

owned this perception since a very young age and it has helped me to understand how I came to the place in my life where it was time, excuse the pun, to write this book. Everything that has happened to me lead me to this moment. My goal is to share my knowledge of time management and life management with you. Mind you, I still have more to learn. But, I owe it to my life lessons and experiences that CEO's, Presidents and other management of companies, from some of the largest corporations in the world to some of the smaller but growing corporations across our nation, have been coming to me for the past seven years for advice on how to market their products and services and then, once they get to know me, ask me to help them operate their companies more effectively. In addition to sharing marketing and advertising advice, which is my first love and was initially the primary service offered by my company, I've had the privilege of working with over 50 companies on issues surrounding employee retention, efficiency, customer service, time management, goal setting, long-term strategic planning, internal communication, redefining internal corporate atmospheres, and sales training. Some of the most successful people in the business world whom I've had the privilege to meet and work with tell me I have a gift...and this gift I now package as Ample Time and share with you. Ample Time addresses more than just time management; it addresses life management and touches on all of the issues from which I've worked with companies over these past few years. If you are missing but one of the simple principles of the Ample Time System in your life, in the way you manage your company, in the way you approach your staff or your customers, in the way you relate to your family and friends...then you are missing out on having Ample Time in your days and a truly abundant life.

As this book comes to its fruition, I thank God for the opportunity to share it with you and pray that you will find the principles within to

be as helpful as my customers, friends, and family have told me it has been for them. Like I said at the beginning of this introduction, you already know the principles of time management and life management. You know how to have Ample Time in your life and accomplish all of your goals. I'm just your little reminder of how to make it actually happen…starting today.

Question:

"How are you today?"

Answer:

"Busy!"

CHAPTER ONE

NOT NOW...I'M TOO ~~BUSY~~ IMPORTANT!

If you visit a library or bookstore or search the world wide web, you will find that there are an unlimited number of books on the subject of Time Management. I would guess there are about 20-30 well-known Time Management systems out there that people consider the "tried and true" standards. Many of them are from the 1980's when self-help books started to gain popularity and our society seemed to go into full gear, trying to accomplish things much faster yet maintain a level of quality. Some books claim that the acceptance of computers into the workplace and home has caused us to work in a faster mode. Some books point to the inventions of microwave ovens, drive-through eating, and overnight deliveries becoming our expected "norm" as the reason we want everything to come to us so quickly. And just look how the internet has helped us to communicate and accomplish tasks in a more timely manner than ever before! But if you research back in time far enough, several decades before computers and fast food, you'll find that even people like Henry David Thoreau were encouraging our society to slow down and "Simplify, simplify!" There are books that suggest we dispose of all our technology and go back to the old way of doing things for a more fulfilling life. Obvious to most people, this is not the answer. We shouldn't have to give up modern conveniences and technology that enable us to enhance our business, edu-

Busy ≠ Important

Busy ≠ Successful

Busy = Busy

cational, and entertainment efforts; instead, we should give up something else entirely.

It is my belief that as a society, we should give up the idea that "being busy" is "being successful" or "being important." If you ask someone today, either in a business situation or casually, "How are you?" their typical response is "Busy!" Even our holidays are jam-packed with running here, running there, rush, rush, rush. In today's world, if you aren't swamped with more than you can handle, you just might be regarded as wasting your time, even if you do consider your workload to be steady. These days, in the eyes of many, steady isn't enough; only busy beyond belief is acceptable. Being busy entails having more on your plate than you can possibly do. For many people, and maybe for you, this kind of excessive expectation has become a norm. We feel like a failure if we are not constantly piling on the demands, either to others or to ourselves. This kind of expectation of busyness has carried over into our leisure time as well. For example, what would be the response of your family members or co-workers if you told them that the only thing you did all weekend was to catch up on your reading or that you spent the holidays lounging around your house and trying out new recipes? For many, you would be regarded as having wasted away your weekend or vacation. You didn't shop, you didn't go out, you didn't clean out the garage, you didn't get caught up on work (yes, we spend our free time working these days, don't we?) etc. You weren't busy enough to please society despite the fact that you spent that time clearing your mind and reducing your stress level, both of which can help you to be even more productive once you return to work.

What I propose is this: instead of "being busy," let's make our goal to "be productive." It's a mental shift. To be productive means that we slow down a bit and plan out how we can make the most of our time

Make the mental shift...

Question:

"How are you today?"

Answer:

"I'm having a very productive day, Thank You."

according to our own priorities, not the priorities of our society. If you think about it, being busy is just that…being busy, killing time. But being productive implies the quality and accomplishment for which we are all striving. We can learn how to have a productive day without a major investment of time or change of lifestyle by immediately implementing the Ample System of Time. I've used this system and know it works. When things quit working for me and I don't seem to be reaching my goals, I turn back to the Ample System and realize that I've strayed from the principles that I know help me to make the most of everyday. Finally, I've seen the Ample System of Time change the lives of those who felt that their time was "out of control" based on the expectations they placed on themselves or the expectations that society placed on them in reference to their career and family. Don't you agree it's about time that we become less busy and more productive? If "busy" means "important" to you in your own mind right now, losing this perception is going to be a challenge. But I promise, if you can change your goal from enjoying "being busy" to enjoying "being productive," you will accomplish more of your goals.

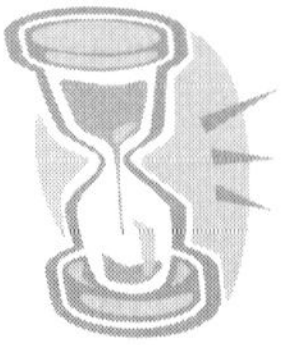

No time for time management?

CHAPTER TWO

WHY CAN'T WE GET IT RIGHT?

It was first during my research for a calendar project that I began my efforts to see varying approaches to making more effective use of time. I attended workshops offered by time management experts and classes offered by our local universities and colleges. I worked with personal coaches and read every book I could find on the subject. Like many scientists who are searching for the cure to one disease, only to discover a cure for another in the process, I was actively seeking the most effective time management system for myself only to find something much more interesting. I was seeing the same people over and over attending time management workshops, working with the personal coaches I was working with, and attending the same time management classes I was attending. I began to ask questions and found out that they weren't implementing any of the time management concepts we were learning. (Come to think of it, neither was I.) It was explained to me that the time management concepts that were presented were too complicated and required a huge investment of time to understand. I had to concur. Where were we going to get more time to learn how to have more time? There already weren't enough hours in the day to get everything done that was expected of us. Upon this realization, I began questioning everyone I met about time management classes they had taken, personal coaches they had used, etc. I

The Ample System of Time

discovered that the majority of people that attend a time management class walk out the door of the classroom and never implement any of the suggestions they just learned to encourage better time management. Have you done that very thing yourself? This is how my mission for a new system began. Somehow, I needed to find out what did work for people, the bottom line on time management techniques, and share it with people in such a way that they could easily memorize it and carry it with them always. That has been my goal throughout the development of Ample Time.

The Ample System of Time is an easy to remember, easy to implement system that allows to you immediately improve your life. It addresses absolutely every life challenge of time management and other challenges as well... I have yet to find any situation that cannot be addressed with the Ample System of Time.

So let's begin. I would like you to take a few minutes and make a list of three to five of the main reasons that the management of your time is a challenge for you. Why are you always so busy and never as productive as you would like to be? What are your personal time challengers? In my workshops, to get people started, I sometimes pose this statement to the participants; fill in the blank:

"If only ______________________________ , I could get a lot more done."

1. ____

____

2. ____

____

3. ____

____

4. ______________________________________

____

5. ______________________________________

____

In the next chapter, you'll be referencing the reasons listed above for your less than perfect time management and more importantly, you'll develop solutions to better managing your time ... starting right now!

Poor time management = a less fulfilling life

CHAPTER THREE

EXTERNAL FACTORS OF TIME MANAGEMENT

As I mentioned in the previous chapter, one doesn't just set out to develop a new system of managing time. At least, I didn't. I actually began learning, and then teaching, other people's systems and when I found out they didn't work, I eventually arrived at the development of the Ample Time System to help myself and others with time management. One thing that is true of many systems, and definitely true of Ample Time, is that we must first identify what is causing us to be time-challenged.

You've now listed at least five reasons for feeling like you are not managing your time well. Let's see if any of your reasons match the most common reasons found in our workshops:

REASONS FOR POOR TIME MANAGEMENT

- If only there were less interruptions during the day (phone calls, unscheduled visits, etc.), I could get more done.
- If only my boss was more organized and helped me to prioritize tasks, I could get more done.
- If only my children weren't sick or didn't have so many activities, I could get more done.
- If only my co-workers were more dependable to do their jobs, I could get more done. (i.e. What my company expects me to do does not match the job description I was given when I

External factors of
time management
never go away.

started my job and I've become very frustrated.)

These reasons for poor time management are common for many of us and I call these reasons EXTERNAL FACTORS OF TIME MANAGEMENT. External means that the root of these particular time management challenges do not seem to stem from your own actions. They are somewhat, if not totally, beyond your control. Or so you think. Beyond your control or not, they are frustrating! There are many more reasons for poor time management that you may have listed. I am only presenting the five most common External Factors of poor time management. External Factors of time management are time-challengers that are most likely not caused by anything you are doing. External factors of time management are "out of your control" and that is why it is so hard to deal with them.

I want tell you something about the External Factors of Time Management but before I do, I want you to take a couple of deep breaths and prepare yourself. When people first hear this news, they are usually disappointed. External Factors of Time Management never go away. They are always with us and no matter what we do to try and change our External Factors, we will either fail or a new, and usually more challenging, External Factor will take its place. We'll always have new reasons to exclaim "There's just not enough time in my day!" Although it's a bit depressing, I want you to get it over with now and realize:

EXTERNAL FACTORS THAT WILL PROBABLY NEVER CHANGE A.K.A. "BAD NEWS"

- There will always be unplanned interruptions in your day.
- Your boss will probably always drop new tasks on you and not always in the most organized manner if that is his/her natural way.
- Your children will always have times they are sick and need

Internal factors of
time management
can help
us deal with
external factors of
time management.

extra attention or a chauffeur service for their activities.

- Your co-workers may not ever gain dependable habits therefore placing extra responsibility on you.
- And finally, as seems to be the case for most everyone these days, you will almost always be expected to do more than your original job description stated.

But take heart…this is the only tough news you'll find in this book. The most exciting news is that there are also INTERNAL FACTORS OF TIME MANAGEMENT. Internal Factors of Time Management and how we address them can assist us to better deal with External Factors of Time Management. Let's take a look at some additional time-challengers that stem from internal factors.

Internal factors of
time management
make up the
Ample System
of Time.

CHAPTER FOUR

INTERNAL FACTORS OF TIME MANAGEMENT

Let's look at your list of reasons for not getting as much done as you would want to get done. Did you include only External Factors of Time Management or did your list also include INTERNAL FACTORS THAT AFFECT TIME MANAGEMENT? Now that you know their are two kinds of factors that affect the way you spend your time, let's define each:

EXTERNAL FACTOR	INTERNAL FACTOR
I have no control over it.	I have control over it.
Others dictate the outcome.	I dictate the outcome.
It changes without my input.	I control any changes.

Now, let's list some Internal Factors of Time Management that you may have listed to complete the sentence. See if you notice how these factors differ from the ones that were listed in the previous chapter. "If only I ________________________________, I could get a lot more done."

- If only I had more energy, I could get a lot more done.
- If only I enjoyed my work more, I could get a lot more done.
- If only I didn't have to work with these idiots, I could get a lot more done.
- The phone is ringing the second I walk into the office. If only

I had a few moments to plan my day, I could get a lot more done.

- If only I had a system of prioritizing, I could get a lot more done. Everything has top priority and I'm not sure which direction to go first.

Internal Factors, whether you realize it or not, affect you the most. Notice the world "I" after "only" in each of the above mentioned reasons for poor time management. Internal Factors relate to the things you can control and when you are not in control, you lose valuable time. When we listed External Factors, we said, "I" after whatever was giving us the time challenge (usually another person or a situation). Do you see the difference?

The bad news I gave you was that the External Factors of Time Management are here to stay and cannot be changed. Now for the good news. Actually, it's great news! The Internal Factors of Time Management can be evaluated and manipulated in a way to assist you in being more effective with your time and to help you get the most out of your day. The Ample System of Time is based on managing these Internal Factors which, while they cannot actually change your External Factors, can help you view the External Factors in a new light. Now, if you understand the difference between External Factors and Internal Factors of Time Management, we should move on. Let's learn the Ample System of Time.

A = ATTITUDE

M = MAKING TIME FOR YOURSELF (EVERYDAY!)

P = PRIORITIZING - A SYSTEM THAT WORKS

L = LIVING IN & LIVING FOR THE MOMENT

E = EXERCISING & EATING RIGHT

CHAPTER FIVE

HOW TO HAVE AMPLE TIME IN YOUR DAY

As I mentioned in the Introduction of this book, I had been working on a "bottom line" time management system for months but was unsure what to call it. The AMPLE name came to me effortlessly one morning in church. This explains a big part of the reason I passionately believe I am supposed to share the Ample Time System with as many people as possible. And, this belief is reaffirmed every time someone mentions to me that the Ample System has given them an easy way to remember a combination of common sense tactics to making the most of life, handling internal and external factors of poor time management and providing a plan for living life with no regrets. Perhaps having Ample Time on earth is part of God's plan for all of us.

The key to having Ample Time in your day is listed below and within the remaining chapters of this book, we'll discuss each of these aspects of Ample Time:

A = ATTITUDE
M = MAKING TIME FOR YOURSELF (EVERYDAY!)
P = PRIORITIZING - A SYSTEM THAT WORKS
L = LIVING IN & LIVING FOR THE MOMENT
E = EXERCISING & EATING RIGHT

AMPLE TIME IN YOUR DAY!

This is the Ample System of Time Management and within its

The
Ample System
of Time
Works!

beautiful simplicity are probably some complex, psychological reasons why it works. I'vc always been interested in psychological reasons for why people respond to certain situations but I haven't spent much time trying to figure out those reasons within Ample Time. While I've always had a passion for studying psychology, I didn't ask psychologists to evaluate this system and tell me why it works. I just asked friends, clients, co-workers, students and other people I met to try it and see if it works. And they tell me it does! Who has time to worry about the reasons why it works? Not us...we are bottom line people and we just need to know the system and get the ball rolling, right? Besides, I think one of the best benefits of the Ample Time System is the fact that you will have a greater appreciation for life and for the lives of others when you implement the principles within Ample Time...and who needs to know why that happens...it's just good that it does.

But here's something we should note; our psychological state has plenty to do with our physical state. Psychologists and medical doctors are researching the link between the mental and the physical aspects of our lives and proving a connection. Poor time management equals stress and a proven physical result of too much stress can be found every Sunday in your local newspaper's obituary section. You may be reading words like "heart attack" and "stroke" but undeniably, some of the obituaries could accurately describe the deceased in the following manner related to Ample Time. Sound like anyone you know? Maybe even yourself?

ATTITUDE

_________ died of a bad attitude. He/She could never see the good in anything and eventually, a poor attitude affected his/her health. (P.S. It was a very small funeral as no one really enjoyed being around their

RIP
Died of
"Toxic
Tension"

bad attitude and most people were glad to see them go!)

MAKE TIME FOR YOURSELF (EVERYDAY!)

_________ died because he/she never made time for him/herself and ran him/herself ragged. He/She didn't take care of #1 because they were always busy doing for others and forgetting to make time for themselves. Eventually, lack of time for his/herself affected his/her health. Those people that depended so greatly on this person would've rather done more for themselves and had this person around even longer in their lives.

PRIORITIZING - A SYSTEM THAT WORKS

_________ died with a few regrets because they didn't prioritize based on their goals. Instead, they hopped from project to project and never had enough time for the most important values or people in their life... "someday" was the motto. "No" was not in their vocabulary. Sadly, someday was yesterday for this person.

LIVING IN AND LIVING FOR THE MOMENT

_________ died after living life in a blurred series of events without ever really focusing on TODAY! There was no living in and living for the moment for this poor soul. Only plenty of activity with his/her mind focused on what was happening next instead of what was going on in the present moment. Surely this was a stressful way to live and eventually, this affected his/her health.

EATING & EXERCISING RIGHT

_________ died after being so "busy" (i.e. "important") that he/she didn't take time to eat in a healthy manner or exercise on a regular basis. "Being healthy" was not their number one goal and therefore,

their other priorities and goals (family, career, etc.) suffered greatly for it. Eventually, this affected his/her health.

One of the most exciting aspects of the system is that it can be individualized to meet your specific needs. As you'll soon see, some of the aspects of Ample Time you have mastered naturally and you are already doing on a consistent basis. Some aspects of Ample Time won't come naturally to you at all but by focusing on those solutions to your time management challenges, you'll see immediate results. Let's go through the Ample Time System, step by step, right now.

ATTITUDE

M

P

L

E

CHAPTER SIX

ATTITUDE

Admittedly, I'm not the first person to come along and tell you that "Attitude is everything!" In fact, I've heard Zig Ziglar and other famous motivators begin seminars with that very introduction! I'm in good company with motivational speakers, management trainers, personal coaches and even dogs who know the value of a good attitude. Yes, I meant to include dogs. For those of you who consider your pets as family, you will really understand this illustration. What happens when you walk in the door each evening and your dog greets you? Your pampered pet is excited beyond belief to see you, isn't he? No matter what kind of day your dog has had, he is thrilled to see you! Dogs don't always have good days, either. Okay, I'm not really that much of an animal person, but I'm guessing that sometimes dogs have bad days, too. Maybe they don't relate well to the neighbor dogs and cats, maybe they get caught in doggie doors that need more grease, and maybe to save money you gave them a haircut that was way to short! But regardless of what happened in their day, how do they act toward you when you arrive home? That's right. They are always excited to see you. Thus, my reasoning for adding dogs to the list.

Do you act as happy as your dog when you see anyone? Most of us don't. Most of us go to work as if we are exhausted from all of the things that have happened at home, ONLY to go home exhausted from

ATTITUDE
M
P
L
E

all of the things that happened at work. My goodness, look at our society - many of us wake up exhausted! If we aren't careful, one day we will wake up on our last day here on earth and, literally, years will have passed with us living in this cycle of blurred activities. You might have experienced this in the past. Maybe you are living it now. Have we made the most of our time? Have we had a good attitude along the way? Or will we have lived life in such a way that we'll have some regrets?

What if we had a great attitude almost all of the time. I mean, let's forget about "not being a morning person" and smile and greet one another! When it comes to attitude, maybe we should act more like our dogs than ourselves! (Please don't take this illustration too literally!) What if we were excited to go from one place to another, to greet the people we live with and the people we work with as if we're happy to be seeing them? I'm not talking about acting "slap happy" and going overboard, but many times, we barely acknowledge those we work with and especially those we live with, taking everyone for granted. If we don't make this effort to have a great attitude, we will probably one day have regrets for not doing so.

I will never forget my last days of college at Texas Woman's University where I studied marketing and journalism. At TWU, I made some wonderful friends and saw many of the same faces day after day. However, for most of us, friendships were limited and from the beginning of the school year to the end, you had your "group" that you hung out with. The very last days of school, something incredible seemed to happen on campus. Everyone was smiling, greeting, acknowledging anyone they came across, helping them carry boxes and books, etc. primarily because they were so happy to be leaving for the summer break. That wasn't the part that impressed me. What impressed me was that I met more wonderful people those last few

ATTITUDE

M

P

L

E

days than I had the entire year because of everyone's attitude. I met people I later wished I had taken time to know. It was the attitude on the campus that made all the difference that last week. What would our year have been like if we had committed to having a better attitude, cheerfully greeting others, and looking for the positive in every situation all year long? That seemed to be what was happening the last week I spent at TWU, and I've carried some of those memories and that experience with me for over 12 years now. That's a powerful testimony to attitude!

I'll never forget my second wedding anniversary and neither will a dear friend of mine. As my husband and I were trying to decide whether to go to a movie or rent one that evening, our phone rang and we learned that our friend, Lynn, had been checked into the local hospital and was experiencing liver failure. The doctors weren't sure just why. Naturally, we rushed to the hospital to support her in anyway that we could, although there was little that could be done until she underwent some additional tests the next day. Lynn was in liver failure. We were all in shock, especially Lynn, as none of us suspected that anything was medically wrong even though she had been feeling exceptionally tired for many months. As is the case for many people and definitely for Lynn at that time, a very hard worker, owning and operating a business can sometimes take its toll on a person physically. Lynn had incorrectly attributed her tiredness to stress and working long hours. To make a very long story short, and get to my point in telling you about Lynn's experiences, I'll quickly describe what happened to my friend and how it relates to attitude. Lynn was immediately instructed to "get her affairs in order" and was transferred to another hospital that was more specialized in liver transplants, as it was decided this was her only hope of survival. Mind you, there was not a liver waiting for her. As her own liver began to completely fail, the doctors

ATTITUDE

M

P

L

E

chose to induce a coma and place her on life support to slow down her body functions and increase her chances of living should a liver become available. Looking back, I suppose it was just a few weeks that all of this took place although it seemed more like a year. I remember how slowly every hour of each day passed by as we waited for a liver to become available for Lynn. One day, Lynn's doctor told her family and friends that she had about 24 hours left to live unless a liver transplant was performed. Miracle of miracles, in the last few hours, a liver was received by the hospital. The transplant was successful and Lynn returned home months before schedule "feeling better than I have in years," to quote her. It was a happy ending that is, unfortunately, *not* usually the outcome for people who are as sick as Lynn was during that time. Recently, Lynn was sharing with me how her perspective on life has changed since her transplant. She told me how surprised she was at the number of people that kept telling her how sorry they were for what had happened to her this past year *when she truly regarded everything that had happened as one of the greatest gifts she could ever have.* "It wasn't fun at the time, but I'm alive because it happened and have more appreciation for life than I ever could've had before." As I listened to Lynn, I couldn't help thinking how great it is to be around people who have a great attitude and a thankful heart for all of their life experiences, even the experiences that are not ideal situations. Do you surround yourself with people who have a great attitude? Do you have a thankful heart and see the good in all situations? Are you that kind of person?

We've all heard of viewing our glass as "half empty or half full" and the results of doing either. It was one of my heroes, Henry David Thoreau that said, "Things do not change, we change." If I were to have you mention your favorite way of spending your time, and asked you to voice it out loud, versus mentioning your least favorite way of

ATTITUDE

M

P

L

E

spending your time, and voicing that out loud, do you know what we would find? When mentioning your most favorite way of spending time, you'll smile, sit up straight, your words will be clear and your tone of voice will be a little higher in pitch, and you'll excitedly go into detail about how you would spend your time. If you had to mention your least favorite activity to me right now, and maybe even just thinking about it will have the same affect, you will not smile and you might even grimace. Your tone of voice will be of a deeper, sadder nature, and you might even mumble and make faces while mentioning it.

It's all in our attitude. I believe there are true attitudes and false attitudes. It was after I was married and decided to start cooking actual meals instead of snacking 24/7 that I had to change my attitude about cooking. Cooking is my least favorite activity in the world unless it is under my own terms. My own terms mean that I have nothing better to do and no one is rushing me and if I fail at whatever I'm trying to make it is okay to order a pizza. Looking back, my bad attitude affected my time management when it came to cooking, as well as affecting the end result.

It's not that I don't love a great meal, I do. But to take valuable time out of my days to grocery shop for ingredients, research recipes, cook the meal and then clean up the mess, is to me a waste of my valuable time when I could be working on business, writing, reading or doing some other fun activity. I have no natural talent for cooking and I totally blame this on my mother! My duties growing up were to set the table and to clear the table...what comes in between those two activities was left up to her. As you might imagine, this presented a problem for me later in life. So I've avoided cooking at all costs most of my life. That's how most people feel about their least favorite activity...like they could be better spending their time in some other manner. But what would happen if I took my least favorite activity and

ATTITUDE

M

P

L

E

regarded it with the same enthusiasm as I do my most favorite activities. Is my attitude toward cooking really justified or is it just that, an attitude. I'm guessing that cooking is regarded as a highly rewarding activity by enough people that four of the approximately forty cable channels on my television feature non-stop cooking shows. So while I'm not alone, I am certainly in the minority with this attitude toward cooking. Because of my many business activities, I used to believe that I really didn't have the time to cook. Not only was this a false attitude, but it was a dangerous belief. Cooking involves fire and in my rush to heat up taco shells (on broil) one evening, I witnessed the taco shells burst into flames right before my eyes. Also, when rushing, note that baked potatoes explode when cooked too quickly and not poked with holes. And, in one of my greatest efforts to save time when I still had a false attitude about cooking and certainly didn't have time to read the directions to instant potatoes, I put real potatoes in with the flake mixture and created a goo that I've rarely seen even in art projects. How many of us carry around false attitudes such as mine was about cooking? My true attitude about cooking, the reality, is that with a little planning, cooking not only saves us money, but it is also gratifying to us to eat a meal that is delicious and healthy! I thought I didn't have time to cook. But the reality, or true attitude, is that if there are 168 hours in each week, and there are, I can plan time to cook.

There are many solutions to a time-challenger once a true attitude is developed. I could devise a plan for myself that would allow me to take only six or seven hours total per week to do the grocery shopping and cook the meals. On the weekend, I can leisurely plan out the week's meals for the entire week, do the shopping and have everything I need for our meals all week long. If I want to try a new recipe, I can bake it on the weekend and freeze it for re-heating during the week. I don't wait to try the new recipe, have it not turn out the way I expect-

ATTITUDE

M

P

L

E

ed, and feel as if dinner is a failure and I've wasted my valuable time! Many times, when I cook, I play my favorite music to make the time go by faster and be more enjoyable. Once I had a true attitude about cooking, it was easy to see the reality in my situation and handle it. Having this true and optimistic attitude about cooking has worked so well for me that I've since moved on to host holiday celebrations for our family and resisted the desire to have them catered! All because I changed my attitude about cooking and now it doesn't seem like an overwhelming and unappealing task. Attitude is everything. Cooking is not a good thing or a bad thing until we view it as such. Now, why have I spent so much time to illustrate how I overcame a dislike of my own? It all relates to attitude. You are probably thinking, "Man, this woman has problems if she can't whip together a meal!" Well, that is probably what many people would think about whatever your false attitude(s) might be. You have false attitudes about many things. Did you know that? It's like green beans. Yes, another food story. Green beans don't taste bad or good...they just taste like green beans. WE decide whether or not we like them. We decide if green beans taste good or bad. That's attitude. And attitude can make a huge difference in how we manage and enjoy our time. How many tasks are you doing that with? *More importantly, how many people are you doing that with?*

About a year ago, I taught a class where a participant pulled me aside at the break and told me that everyone she works with is a complete imbecile and that is why she can't get anything done. Obviously, she had an External Factor of Time Management (imbeciles to work with) that was affecting her time. Or was it her attitude? We began discussing specifics about her co-workers and after a few minutes, it became clear that her co-workers were not ignorant at all; in fact, most of them held very high collegiate degrees in their field. What was in

ATTITUDE

M

P

L

E

fact happening was that her co-workers completed tasks in a different manner than she did (External Factor) and this affected her time and she began to develop a negative attitude towards them (Internal Factor) and this just made her situation even worse. She agreed to try a little experiment with me. Everyday for one week she said to herself, at least 10 times on her drive to work, "I enjoy working with intelligent people and I will show them my respect." In addition to this "co-worker mantra" as we called it, she did follow through with our plans and showed all of her co-workers her utmost respect. This meant no more snide remarks, rolling of the eyes, or explanations on why her way was better when they had a question or comment. She wasn't even allowed to think a negative thought about her co-workers or to go home and vent to her husband about how she did not enjoy working with these people. I didn't expect to hear from her until the end of the week. After all, going from calling her co-workers "imbeciles" to "intelligent people" could take some time. But because of her dedication to changing the situation and her understanding of the Ample System of Time and the importance of attitude within it, she made a difference in her situation almost immediately. She called me on day two of our little Ample Time experiment and admitted that her attitude towards them had more to do with the situation than she realized. The results she experienced by the end of the week were phenomenal! She expressed that she was actually looking forward to going to work and she shook at the thought of having almost spent the next 10-20 years with her bad attitude. The time she spent at work was no longer a drag. Her time was spent accomplishing goals with competent co-workers and just by her change in attitude, time at work was no longer wasted. Her personal time with her husband improved, too, because that time was no longer spent complaining about her day. I want you to note what I said about her change in attitude…for we are always wishing others would

ATTITUDE
M
P
L
E

change and thinking that is the only way for a situation to become better. Many times, if WE are unhappy, WE need to change. We = Me in this case. This is an example of an Internal Factor of Time Management being able to affect External Factors with a better outcome for everyone.

Another great example of how to improve Time Management is to implement within your company or organization what I simply call "The No Negativity Rule." How much time are you (or if you are managing a company ask yourself if this is true of your employees) spending in lieu of work to "pass along information that may or may not be true about other co-workers, customers or management?" I'm talking about opinions…gossip! This eats up more time in companies than any other activity I've ever witnessed, of this I am convinced. And for those corporations who have implemented the "No Negativity Rule," the results have been dramatic. Productivity soars. People who didn't have enough time to do everything suddenly have time to do even more. You can't implement such a program unless you are going to be serious about it. I recommend a "three-strikes and you're out" policy that gives employees two warnings and one "good-bye, nice knowing you, here's a box for your things" meeting. This kind of a rule can weed out those people who are not serious about your company's goals and allows you to keep the people who can adopt the true attitude that most corporations want their employees to exhibit. Whether it is for yourself or for your company, making a rule like the "No Negativity Rule" can immediately position you and your company with a truly good attitude internally and externally. Everyone - your customers, potential customers, and employees - will notice and enjoy a company-wide positive attitude. Best of all, your customers will want to work with you even more.

Make a list of all of the things about which you could have a bet-

A TTITUDE

M

P

L

E

ter attitude. Find the solution and make up your own "attitude mantra" to remind yourself about your new attitude. Maybe you should place your attitude mantra on your bathroom mirror so you see it in the morning, or maybe on your visor in the car so you can say it over and over again on your way to work. Try it for just one week and make note of the differences your new attitude makes in your life and best of all, the way you manage your time.

MY CURRENT "FALSE ATTITUDES"	MY NEW "TRUE ATTITUDES"
____________________	____________________
____________________	____________________
____________________	____________________
____________________	____________________
____________________	____________________
____________________	____________________

I want to end this chapter with the most powerful example I've personally experienced of what can happen with a good attitude. It was in third grade that I first remember someone commenting on my attitude. I remember it as clear as if it happened yesterday. Greenwood Elementary, out by the jungle gym, right after lunch…my friend Julie said, "Lorie, you are always smiling." It still strikes me as odd that a comment like that would come from a third-grader, or that I would remember it, but I do. Thanks to a mother that sang "Mary Sunshine" songs every morning to wake my brother and me and didn't allow a bad attitude regardless of any hard times or circumstances, I have been smiling all of my life! Incorporating a good attitude in my life has paid

ATTITUDE

M

P

L

E

off in ways that I'll never know but in 1993, I had the unusual opportunity to realize what a good attitude created for me. One of my first clients was the world's largest retailer and having worked with this company for over a year, I sent out a survey asking for comments on how I might improve my service to them. One of the questions was, "The main reason we use your services over other companies offering the same service is…" and there was a space for them to write down any reason that came to their mind. This was not multiple choice and this questionnaire went to about 10 different people within the company with whom I had developed relationships. Anxiously awaiting their responses, I wondered if they would list our prompt response to their phone calls to answer questions, our one-two day turn on "hot projects," our creative abilities and graphic design talents, our organization charts on each project that meant they never had to wonder what we were doing for them at any given time…I couldn't wait to find out which service they appreciated the most. Would you believe that 9 of the 10 responses came back with comments like "Lorie's great attitude," "I always look forward to Lorie's cheerful ways," "Lorie brightens up our office with her visits." I tell you this not to brag but because it was such a telling moment to me on the value of a good attitude. Upon receiving their responses, I initially felt deflated. Understand, I had worked so hard on offering excellent creative services, making sure our invoices always matched our quotes, returning phone calls the minute I was paged…I was a little disappointed they didn't seem to notice any of those things about our company. As for the attitude, I was just being myself. Then it occurred to me that any company that wants to can offer all of those services I was offering…but apparently, it was a good attitude that gave my company an edge and helped us to obtain and maintain that account. That was the moment when I realized the value of having implemented a good attitude toward myself

ATTITUDE

M

P

L

E

and others years prior to that moment. I had been forced to do so by my mother when I was young, and eventually, it took hold of me and became part of my true personality. I automatically had a good attitude and didn't even know it to be different from others until that moment. But apparently, others were noticing! Powerful, isn't it? I've had the opportunity to see the atmosphere of companies change when CEO's and their management team improve their attitude and share that positive vibe with others in their organization. Yes, a good attitude, it is powerful! It's never too late to get started on making your attitude an even more positive one, either. P.S. Thanks, Mom.

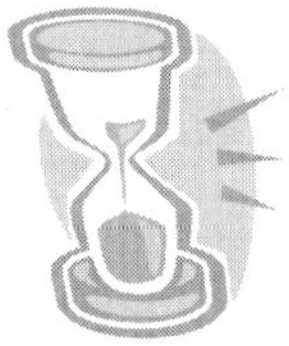

How is my

"Attitude Gauge?"

A

MAKE TIME FOR YOURSELF (EVERYDAY!)

P

L

E

CHAPTER SEVEN

MAKE TIME FOR YOURSELF (EVERYDAY!)

Understanding that a good majority of society lives in a very fast paced world, it's not unheard of that many people's version of a "normal" day might look like this: hit the alarm clock several times, sprint to the shower to try to make up for not getting up on time, jump out, get dressed, feed and dress kids, rush to work in traffic, cram down breakfast on the way to work, be greeted with demands and a pile of phone messages, attend meetings, attend meetings, attend even more meetings, run through a drive-through at lunch, rush to pick up the kids, arrive late to cub scouts, eat dinner late, run the kids through the bathing cycle, check homework as quickly as possible, read part of a bedtime story to the kids, stick another load of laundry in the washer, respond quickly to personal email, feed the pets, watch the news, flop into bed and fall asleep exhausted only to repeat this cycle the very next day. Sound familiar? You may be able to change the events of the day in the description above and still be saying to yourself, "Yes, that's about how my life goes, too!" If life seems as if it's a whirlwind tour to you, the question then becomes how to stop this cycle and make the most of every day. Look no further than this chapter of the Ample System of Time. You are a victim of "busyness," but not for long.

Making time for yourself isn't a new concept, it's just one that many of us have forgotten. In order not to go from activity to activity

A

MAKE TIME FOR YOURSELF (EVERYDAY!)

P

L

E

in a mad rush, we must have a few moments to reflect and plan and decide what we want to accomplish. This step of the Ample System of Time is so very important for accomplishing goals and making the most of every moment. If you can spend just a few moments concentrating on how you want your day to go, and reflecting on the day you just experienced, you'll force yourself to slow down a bit and break your hectic activity cycle. Through this process, you will see instant changes in how you better manage your time! Why is this? Well, you may have heard famous motivational speakers telling you how important it is to meditate to see the outcome of how you would like your day to go. In addition, you've probably heard the famous saying, "Without a plan, you are planning to fail." Well, this advice is exactly right. The Ample System of Time suggests that everyday we plan how we would like the day or week or weekend to go. I'm not talking about writing down your plan and checking off activities one by one. That's a "to do" list. What I'm suggesting is a more broad sense of how you want to feel at the end of the day: productive, happy, enjoyable, accomplished, satisfied, relaxed, under control. Pick the word that is right for describing yourself and make it your own.

For a person who is living life from activity to activity in a big rush, words such as confident, controlled, relaxed, and positive represent the opposite of how they usually feel at the end of the day. If you think it is impossible for you to have even five or ten minutes of planning time and reflecting time each day, then I suggest you take your planning time during some activity that is spent alone. Examples of this kind of time might be getting up five minutes before everyone else in the house is awake, planning and reflecting during a hot, relaxing bath (my personal favorite) or while you take that ever-cheerful dog with the great attitude for a walk around the block. Notice I said planning and reflecting...this is also your time to review the day, admit

A

MAKE TIME FOR YOURSELF (EVERYDAY!)

P

L

E

mistakes and rehearse in your mind how you'll not make the same mistakes again. Likewise, when you've done something wonderful during your day, you should compliment yourself mentally…sometimes, you'll be the only one who gives yourself the recognition you deserve! Why is this? Because everyone else is so busy, of course. If you will take this time for yourself to focus on the day that has just passed and plan for your tomorrow, you'll immediately notice how much smoother things will go and how you will feel less rushed. This is because you are taking time to think about, plan for, improve and appreciate your moments in time. When things start to feel out of hand, and you need to take control of how your time is being spent, remember to Make Time For Yourself…Everyday!

TIMES OF THE DAY SET ASIDE FOR MY PLANS, THOUGHTS & REFLECTIONS…

How full is my

gauge labeled

"Make Time for Myself

(Everyday!)?"

A

M

PRIORITIZING - A SYSTEM THAT WORKS

L

E

CHAPTER EIGHT

PRIORITIZING - A SYSTEM THAT WORKS

There are many buzzwords in the time management industry and one of the biggest ones is "procrastination." Procrastination occurs when one puts off doing a task or chore or activity because they dread having to do it. The true irony occurs when we procrastinate for a greater amount of time than it takes to do whatever we are procrastinating about. Sad but true, much of our time is spent thinking up ways to avoid doing activities we don't want to do. Procrastination is just one example of human behavior that relates to time management that can be solved through the Ample System of Time. Prioritizing is a big key to effectively disarming procrastination tactics and having Ample Time in your day.

The problem for many people is that they haven't taken the time to really list their priorities. Or, when they do take the time, they have so many priorities that it overwhelms them as to which are the most important priorities on their list. Having read many books on the subject of prioritizing, researched the best ways of making "to do" lists and interviewed numerous participants of my workshop on which is the best way to prioritize, I find it no wonder that many people just toss up their hands and forego prioritizing altogether. The current methods proposed by many time management experts assume that each person has the same personality characteristics and habits. Some methods get

PRIORITIZING - A SYSTEM THAT WORKS

very technical on the proper way to make "how to" lists, how to determine if a task is truly completed, what to do with tasks that are halfway completed, etc. It would take most of us half a day just to set up such a system and hours each day to maintain it. What we will learn in this chapter is very simple and best of all, my students and workshop participants with different characteristics and habits tell me it works. If you are the kind of person who tends to hop from activity to activity and never really accomplishes any project, (some people call this "putting out fires"), this chapter will be especially meaningful to you.

First things first. We must have overall priorities in life to even know where we should spend our time. If you are following the Ample System of Time, you are only going to have three of what I call "Big Goal Priorities" in life. You could also consider these "Big Goal Priorities" equivalent to "Your Purpose in Life." For some of you, this will be a difficult exercise. As one women proclaimed during a workshop, "You don't understand, I have about 20 or 30 priorities in my life..." She felt unfulfilled and didn't know why. We went on to discuss her priorities and found out that in addition to her family, her commitment to religion, and her career, she also felt it was a priority to donate as much time as possible to being a board member of five different charities, remain an active voice in her state legislature, and keep up on issues involving world peace. And finally, she felt it was a priority to encourage everyone she met to be as environmentally responsible as she tried to be. No wait, she also added that she was having problems with being successful in her sales career, although she loved the flexibility her job gave her to do all of these other activities. This was definitely a ~~busy~~ important woman. Maybe you can see why this woman felt that she didn't get much accomplished...but she couldn't understand why she wasn't accomplishing her goals and getting more done. After all, she was doing so much! Finally, in the workshop, this

A

M

PRIORITIZING - A SYSTEM THAT WORKS

L

E

woman realized that having good intentions is very different than having a priority when you use the Ample System of Time.

The Webster's Dictionary defines a priority as "precedence in order, privilege, etc." For our purposes, let's define "Big Goal Priorities" or "privileges" as something you get to work towards each day of your life until the day you die. We are talking about something that is so important to you that you will work on behalf of this goal for your entire life. Now, what are your top three (and only three) "Big Goal Priorities" in life? Remember, we are talking about your purpose in life. You might need to take some time out here to really think about what you want to list as your priorities.

What? You say you can't name just three. Let's discuss this a bit further just to make sure you understand what I mean by "Big Goal Priorities." Big Goal Priorities are deep in your core and they will rarely, if ever, change. If it was possible from a mental standpoint, these are the priorities that you would've stamped yourself with when you were born, to carry them with you throughout your life with little or no change. They are specific in nature and regardless of changes in your personal circumstances, (marriage, kids, career) the same three "Big Goal Priorities" will remain with you most of your life.

The less specific you are, the easier this exercise may be; however, the more specific you are about your priorities, the better results you'll get in managing your time because you will have a very clear idea of how you should relate your actions to your goals (i.e. spending your time). Now list your priorities and then we'll go into a little more discussion to see if you are going to keep these priorities or change them based on our discussion.

A

M

PRIORITIZING - A SYSTEM THAT WORKS

L

E

BIG GOAL PRIORITIES - GENERAL

1. ____

____

2. ____

____

3. ____

____

Did you list your three priorities? I'm going to give you an example of what I tend to see time and time (excuse the pun) again as I conduct this exercise in workshops. Many people list their top three priorities as 1) religion; 2) family; and 3) career, and usually in this order. Funny thing is, as we begin to explore how they are presently spending their time, their actions usually do not reflect what they claim as their priorities. So, I ask them to get very specific with their "Big Goal Priorities." For example, if religion is listed as a priority, they could relate it to a lot of different activities. Do they place importance in their lives on attending worship services, assisting those people who are less fortunate, meditating or praying on a regular basis, or making sure their actions are pure and reflect their religious beliefs? If family is important to them, do they mean the family into which they were born, the family they have or plan to create, an extended family, an adopted family, their church family, etc. What about their career is so important to them that they list it as a priority? Is this a priority to enable one to pay bills or is this a totally different priority in which one will gain fame or recognition or fulfill a personal achievement? Career as a "Big Goal Priority" could also be listed as a means for leaving wealth to future generations of one's family or to help an organization

A

M

PRIORITIZING - A SYSTEM THAT WORKS

L

E

or company further reach their goals through your years of dedication, intelligence and hard work in exchange for a comfortable retirement. Do you see how these descriptions differ from just listing a one-to-three word priority which we so often do? Now, with this explanation in mind, re-list your top three "Big Goal Priorities."

BIG GOAL PRIORITIES - "MY PURPOSE IN LIFE"

1. ____

____

2. ____

____

3. ____

____

Having very specific "Big Goal Priorities" allows you to determine the best way for you to spend your time throughout the day. It automatically puts many of your activities that you are "squeezing in" right now on the back burner for times when you are ahead of the game and can do those "good intentions." How many of your daily activities are helping you work toward "good intentions" instead of "Big Goal Priorities?" Let's try and list them now. On a separate sheet of paper, make a list of everything you did this week so far and evaluate it to get the best picture of how you are currently spending your time versus how you will begin to spend it based on your "Big Goal Priorities." After listing your activities, highlight those events which help you reach your "Big Goal Priorities" and re-evaluate the activities that don't lead you in the direction of your goals. Many times, these activities can include organizational meetings for work or your neighbor-

A

M

PRIORITIZING - A SYSTEM THAT WORKS

L

E

hood, charity efforts, unplanned or unscheduled meetings or lunches with suppliers and vendors with whom you have little intention of doing business, etc.

Years ago, I heard a talk show host, Oprah Winfrey, describe how to stay focused on what is most important in one's life. She explained how she was very busy with activities and didn't have time to do everything she wanted to do, including taking better care of herself. She was running herself ragged trying to keep everyone else happy. Finally, Ms. Winfrey learned to "just say no" to the assumed priorities (really just good intentions) that weren't her most important activities based on her priorities. From this example we learn that if your daily activities are not directly helping you with your "Big Goal Priorities," perhaps you should "just say no!" You should be spending at least one-third of your day on Big Goal Priorities. Since one-third of your day is equal to eight hours, this is an important exercise. The reality is that you probably spend more than eight hours on your true "Big Goal Priorities."

Okay, so you've listed your big priorities, but what about all of the "less important but still important" priorities or "good intentions" that you struggled with earlier in this chapter? I'm talking about the less important, but still important, priorities you think must be listed somewhere! Put these goals or important activities on another list called "Good Intentions." These might include spending time with friends, learning a new language, traveling around the world, volunteering time on a charitable project, etc. Your "Good Intentions" are priorities that you would like to be involved with or do but may not have the time to complete, depending on how closely tied they are with your 'Big Priorities." You should be just as specific with your "Good Intentions" as you were with your "Big Goal Priorities." And, unlike "Big Goal Priorities," your "Good Intentions" can change over time. However,

Are you so busy
that you tell everyone
how busy you are?

1) This wastes more of your time.

2) This wastes other people's time.

note that you are only going to spend no more than three hours per day on these activities and therefore, you must choose just two of them to focus on:

"GOOD INTENTIONS"

1. ____

____

2. ____

____

Do you like to fail? Of course not, nobody does. But how many times do we set ourselves up to fail by trying to attach ourselves to every good idea or good cause that comes our way. We want to be a part of everything and we claim the good causes and good intentions to be our priorities. For safety's sake in a life mission of no regrets, choose a limited number of causes or good intentions in which to be involved. You'll see more successes in reaching your good intentions when you are focused on just a few goals that you can really accomplish instead of trying to be a part of many opportunities.

If you're stuck and can't list your Big Goal Priorities and your Good Intentions, then make a list of everything you did each day this past week. Your actions will speak your Big Goal Priorities for you. Do they match what your true Big Goal Priorities should be? Do you need to make a shift in your actions to meet your true Big Goal Priorities? Are you spinning your wheels with your time or moving in the direction of your goals?

Now for those of you who are still in love with the concept of being "busy" and equating "busy" to "successful" or "busy" to "important" this has not been a very easy chapter to complete. If it makes you

A

M

PRIORITIZING - A SYSTEM THAT WORKS

L

E

feel any better, I personally know what it is like to make the transition from "busy" to "productive." Perhaps you act as I did at one time, not only spending your time actually being "busy", but when you have a few spare moments, you innocently profess to others just how very busy you are and take up their valuable time as well. Take it from me, productive is better than busy. By getting specific about your goals and your intentions and making sure your actions follow suit, you'll soon find that you get more out of your moments! Don't wait to wake up to this reality until something really major happens in your life to make you realize that your time is very valuable (i.e., health problems, an accident, realizing you can't fulfill a goal due to a specific limitation, etc.). How you spend your time today and everyday after quite literally determines the value of your life.

Now, I really don't want to be the one to tell you how to make a "to do" list but in literally every workshop I've hosted, this question comes up. So, the final piece of advice that I have to offer in reference to prioritizing relates to maintaining a "to do" list that many of us use on a daily basis. I think a "to do" list is a necessity, but the form you choose for it and the manner in which you maintain it is up to you...with just one rule. Choose only three things on your daily "to do" list that you WILL MOST DEFINITELY, NO EXCUSES, ACCOMPLISH each day and make sure that you try and complete these tasks as early as possible in your day. By choosing just three items on your list, and making sure that you complete them each day, you'll make huge progress in life because at the end of the week, you will have accomplished at least 15 significant tasks. You may say to yourself, "Hey, I get a lot more than three things done everyday!" That's great and hopefully that happens. But for many people in my workshops, we discuss how easy it is for an entire day to pass, hopping from project to project without actually completing anything. Another

A

M

PRIORITIZING - A SYSTEM THAT WORKS

L

E

great "to do" list rule…if you are moving tasks from day to day and the projects just seem to go on throughout the month and never get completed, make it a rule to move a task that you are in control of no more than three days. If the task has been moved from day to day three times, you are procrastinating and this task MUST be completed on the third day.

Prioritizing is a must in all time management systems and the Ample System of Time is no different in this respect. Determine your life priorities and your good intentions and act accordingly…soon you will be making the most of everyday!

MY PRIORITIES

	MONDAY	TUESDAY	WEDNESDAY	THURSDAY	FRIDAY
1.	________	________	________	________	________
2.	________	________	________	________	________
3.	________	________	________	________	________

How full is my

"Priority Gauge?"

Is it too full of good intentions

to accomplish my

"Big Goal Priorities?"

A

M

P

LIVING IN & LIVING FOR THE MOMENT

E

CHAPTER NINE

LIVING IN & LIVING FOR THE MOMENT

In 1993, I opened an advertising agency and a potential customer of a very large company told me how much he loved to play golf. I had been told by some mentors that "more big business deals are done on the golf course than anywhere else," and since I had always wanted to learn how to play golf *and* do big business, I purchased some clubs and took golf lessons. It was during this time that something very interesting happened. Every time that I was in an office meeting with my customer, all he could talk about was how much he wished we could be playing golf that day. When we were actually out on the golf course together, all he could talk about was how he should really be back at the office working. This is a perfect example of someone who was not Living In or Living For The Moment.

Living In & Living For The Moment means staying focused on what is going on in the present moment. It wasn't until I was married and had step-children that I realized something wonderful about children in general. They usually don't worry about what happened yesterday or what is going to happen tomorrow. Children are completely focused on what is happening right now. Children learn new concepts quickly by Living In & Living For The Moment. Children show true appreciation for your attention by Living In & Living For The Moment. Children are extremely creative by Living In & Living For The Moment. Children laugh out loud often by Living In & Living For

FOCUS =

LEARN MORE

LOVE MORE

LAUGH MORE

LIVE MORE

The Moment. It occurred to me one day while watching my step-children and their friends play that by Living In & Living For The Moment, we would all be much better off. While this concept has been realized by countless others before me, until I made that realization for myself, my time was not being spent as effectively as it is now. I believe the same will be true for you.

There is a single-word term for Living In & Living For The Moment - Focus. Focus on what is happening right now and do not let anything distract you from it. Are you talking on the phone at work to a customer or colleague? Put down that pen and paper and listen to what that person is telling you. Is your child trying to have a conversation with you? Stop what you are doing and focus on them for just a few moments. Is your spouse relaying to you what happened to them during their day? Give them your full attention for a little while. By Living In & Living For The Moment, you are giving a great gift to others, but you are receiving an even greater gift. The time you spend focusing on the present moment can lead to learning more, loving more, laughing more…living more! This makes for a more fulfilling life in which you are making the most of your moments. When you are making the most of your moments, you are making the most of your time.

It was during one of my workshops that I was just approaching this segment, "Living In & For The Moment," when a cell phone belonging to one of the participants rang. It was the third time that this had happened during the workshop and for each call that came in, this participant would fumble in her purse to find the phone, "whisper" loudly, "Hello?" (yes, there is such a thing as whispering loudly) then climb over a row of participants to get out the door and into the hall so as not to disturb the rest of us. Right, she wasn't disturbing anyone at all. Then she would climb back across everyone to her seat and sit atten-

A

M

P

LIVING IN & LIVING FOR THE MOMENT

E

tively only to repeat the process with the next call.

Fortunately, this sharp lady was in her seat when we discussed "Living In & Living For The Moment." During our discussion on focusing on the moment, she suddenly realized that if she was spending her time answering calls, running in and out of the classroom, and concentrating on anything but our workshop, she was going to miss out on learning the Ample System of Time. Worst yet, in the process of running in and out of the classroom to take calls, she was disturbing others who were spending their time with us. This woman was intelligent enough to turn off her phone the moment that she came to this realization, and secure enough to share her lesson with us at the end of the workshop. And what a valuable lesson we all shared that day as when we don't focus on the moment, we very often make less of other people's time as well as our own. Think about it. Live In & Live For The Moment!

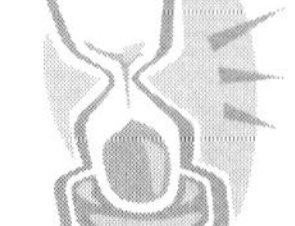

[illegible]

Is my

"Living In The Moment Gauge" full

or am I rushing

through life?

A

M

P

L

EXERCISING & EATING RIGHT

CHAPTER TEN

EXERCISING & EATING RIGHT

I remember thinking in senior high school that I was a little overweight when I went from 95 pounds to 98 pounds right after my junior year. While I don't care to share with you my current weight, I will share with you this much. I wouldn't mind knowing what that "tubby" 98 pounds feels like again, even if for just one day. Back in those junior high and high school days, I could eat anything I wanted and didn't think too much about exercise. A daily routine of walking to classes and after school activities kept me in fairly good shape.

As a nation in general, we are very unhealthy. For most of us, as we grow older, we need to pay more attention to our eating habits and exercise routines. If you think your state of health doesn't have anything to do with the way you manage your time, think again.

Like so many people, to maintain a healthy weight and have plenty of energy is a goal that I constantly strive for, and I seem to have better days than others when it comes to making healthy food choices and exercising often. Choose whatever nutrition and exercise regimen best fits your lifestyle; just make sure that you do choose one. And in my opinion, exercise is imperative in clearing ones mind, as much as it is for burning calories. Remind yourself...exercise is for the body and the mind. So what exactly does diet and exercise have to do with how you manage your time?

A

M

P

L

EXERCISING & EATING RIGHT

Perhaps it was one of my former workshop participants who said it best. When she exercised on a regular basis and didn't eat any "junk" food, she seemed to have less problems at the office and managed her time much more effectively. It just makes sense that if we take care to maintain our health and have more energy, we can get more work done, we will work more effectively with our co-workers and we'll feel better about ourselves. For these reasons and more, we can better manage our time with a proper nutrition and exercise regimen. Our energy level is yet another Internal Factor of Time Management that we can control and put into motion to favor our "Big Goal Priorities" and help us maneuver around the External Factors of Time Management that trouble us. See how it's all connected? I hope one of your "Big Goal Priorities" is to remain healthy throughout your entire life. As I sometimes remind my workshop participants, without health as a top priority, it's hard to take care of our other priorities. Not to mention all of our good intentions! Health is an Internal Factor of Time Management that you can significantly control!

How is your

"Eating & Exercise Gauge?"

Are you feeling

full of health

or running a little empty

on energy?

3

CHAPTER ELEVEN

THE RULE OF THREE

I mention the number three quite often in this book. I've found three to be a number that I can rely on to help me reach my goals. In reference to Ample Time, here's how using the number three can help you make the most of your time beginning right now!

1. Choose three people that you haven't been especially nice to in the past and show them your new, positive attitude towards them.
2. Choose three External Factors of Time Management that are holding you back from effectively managing your time and determine the solution based on the Ample System of Time.
3. Choose three Internal Factors of Time Management that are keeping you from being effective with your time and determine the solution based on the Ample System of Time.
4. Think of three ways that you could Make Time For Yourself Everyday during the day, even if only for a few moments. Choose one way to implement today and begin planning to be "efficient" or "happy" or "productive," etc.
5. Review your three "Big Goal Priorities" and put them down on a paper that you can hang on a computer, mirror or somewhere you'll see it often. These are the goals of your life that never change. You will always have these "Big Goal Priorities."

3

6. Choose three times during the day today when you were not Living In & Living For The Moment and think about how you will change that kind of behavior tomorrow.
7. Choose only healthy foods and let yourself have a "treat" such as dessert just three times per week. Exercise a minimum of three times per week.
8. Read this book three times each year to remind yourself of the Ample System of Time.
9. Review three of your current "false attitudes" and change them into your new "true attitudes."
10. Go out and buy three of the Ample Time books to give to three of your friends or co-workers…okay, maybe I'm getting a little carried away!

The Ample System of Time Works!

CHAPTER TWELVE

WHY THE AMPLE SYSTEM WILL WORK FOR YOU

Like I mentioned in the Introduction of this book, my initial goal wasn't actually to create a system of time management and profess it, but one thing lead to another and as the Ample System of Time was developed I began to share it with others. Unlike the other time management systems I had been teaching when I first started, people were commenting to me on how this, the Ample System of Time, is the first system that had worked for them. It's common sense. It's full of principle's that we have been taught all of our life. It's easy to remember. It's easy to immediately implement. It produces results the moment one begins to use it. It is the answer for anyone who struggles with making the most of their time or anyone who feels their life is a blurred series of activities and they can't gain control. The Ample System of Time maximizes the moments of your day. Life is made up of moments and maximizing the quality of those moments is the challenge we all face everyday.

Make the most of your life! My most sincere wishes to you in being successful and having Ample Time in each and every one of your days.

"Maximize Your Moments!"

Lorie

YOUR NOTES ON THE AMPLE SYSTEM OF TIME

EACH LETTER STANDS FOR WHAT?

A = ______________________________

M = ______________________________

P = ______________________________

L = ______________________________

E = ______________________________

"We seldom value
the importance of
remembering
until we forget."

Lorie

Which part of the Ample System of Time will help you better manage your time in the future?

ATTITUDE

MAKE TIME FOR YOURSELF (EVERYDAY!)

PRIORITIZING - A SYSTEM THAT WORKS

LIVING IN & LIVING FOR THE MOMENT

EXERCISE & EATING RIGHT

What immediate change(s) are you going to make to have better time management?

Personal notes:

THE GOAL